SPACE ELEVATOR OPERATIONS

Book 3 of the Space Elevator 2020 series

Linda Phillips

CONTENTS

Space Elevator Operations
Book 3 in the Space Elevator 2020 series

The concept of a Space Elevator started over a century ago but started to move from science fiction to reality in the 1990s with the promising advent of a super strong material, carbon nanotubes, that theoretically had the strength to build a ribbon into space.

NASA started investigating the concept which led to the publication of the first detailed analysis of whether it was possible, written by Dr Brad Edwards, in a NIAC paper.

Early excitement of an immediate breakthrough, maybe even constructing one by 2010, gave way to the reality of producing carbon nanotubes of the required strength, outside the laboratory, in commercial quantities. As I write, in 2020, we are still awaiting this breakthrough and the project has moved back to the late 20s, maybe the 2030s or 2040s.

But it will happen, and when it does, it will replace rockets as the easier and cheaper way to leave Earth, opening up space travel the way airplanes opened up world travel.

This series examines the technical aspects and how it will be deployed.

In book three, usage of the ribbon is addressed, along with issues that arise out of putting humans in space for prolonged periods; the setup and usage of the two space stations on the ribbon, the Geo Station and the Outer Space Station; plus how these will be used to travel to the Moon and Mars.

1. REFLECTION

In 1957...
no rocket had ever traveled to space.

Yet 12 years later in 1969...
astronauts were walking on the Moon!

Once we've constructed the first space elevator, how do we use it?

In this book, we consider practical design issues for the cable cars intended to travel on the ribbon and related ribbon management issues. We also consider power delivery systems, and a primer on the location in the ocean of the first Elevator Space Port.

Then it's on to the exciting parts for the first astronauts to travel into space on the ribbon: the construction and use of the first two space stations, and how they are used to act as docking ports for our new spaceships which will travel onwards to the Moon, Mars and elsewhere in the solar system.

2. RADIATION IN SPACE

Before we send anyone up the ribbon into space, we have to consider how to manage the dangers of radiation. At the Geo Station, radiation is far worse than at the much lower ISS. If we are considering putting people, workers, astronauts into space for lengthy periods, we have to manage their cumulative exposure.

Radiation in space is a problem, but until you have to think about spending a long time in space, you don't appreciate the severity of the problem.

Where does the atmosphere end, and space begin?

There are astronauts in space all the time, at the ISS, for sure. But the ISS is in LEO a mere 400 km above Earth, which is barely in space. We commonly say space begins 100 km above sea level, where the atmosphere ends, sort of. It actually fades away with altitude, but at 100 km up there, there is hardly any atmosphere to speak of, and the sky becomes the black of space. So the ISS is 300 km above the atmosphere, by the common measure.

Throughout this series, we've adhered to the common definition of space, the Kármán line, at an altitude of 100 km (62 mi) above sea level, by convention. At this altitude, air pressure is 0.006% of sea level pressure, non-existent for most practical purposes.

There are other definitions of the beginning of space. An alternative is 80 km (50 mi) altitude, this being the altitude achieved by the first US astronauts, Alan Shepard and Gus Grissom, on their sub-orbital loop, in 1961. They were called astronauts for achieving this altitude, and even today, NASA still uses this definition, in deciding whether to award someone the title of Astronaut.

Reports National Geographic:

'The Federation Aeronautique Internationale (FAI), which keeps track of standards and records in astronautics and aeronautics, also defines space as beginning a hundred kilometers up. It is, after all, a nice round number.'

'But the Federal Aviation Administration, the U.S. Air Force, NOAA, and NASA generally use 50 miles (80 kilometers) as the boundary, with the Air Force granting astronaut wings to flyers who go higher than this mark. At the same time, NASA Mission Control places the line at 76 miles (122 kilometers), because that is "the point at which atmospheric drag becomes noticeable.'

At 10 km altitude, the air is already too thin for humans to breathe. If you are flying in an airplane at this altitude, a loss of cabin pressure calls for oxygen masks to drop down, so you can breathe. Without these you'd be unconscious in fifteen seconds.

Traces of atmosphere extend to about 1,000 km altitude.

But, above the ISS are the Van Allen Belts, a sort of anti-radiation magnetic region which protects our planet from the full-on blast of space radiation. The Van Allen Belts are from about 600 km to 10,000 km above sea level. In a sense, the belts can be considered a supra-atmosphere level, with true space only commencing above 10,000 km from us.

In one sense, the atmosphere could be regarded as becoming negligible at about 600 km altitude, where the Van Allen Belts replace the atmosphere.

So, in reality, the ISS is not even in true space, for practical purposes, since it orbits underneath the Van Allen Belts and is within the atmosphere, admittedly the thin end of it. But, for consistency, we adhere to the Kármán line at 100 km as the definition of the end of the atmosphere. It's a nice round number. Space engineers understand the atmosphere doesn't hit a roof, it just fades away. The definition of the edge of space has more importance politically and legally. If a low definition is used, 80 km, then a spy satellite orbiting over another country at 81 km altitude is in space, a legal commons, whereas at 79 km it is an invasion of sovereign space.

People have been beyond LEO to the Moon, of course, but for each of them, it was a hurried trip of about ten days duration. Part of the reason was to limit radiation exposure for the astronauts.

So, when we start to consider building a Geo Station, a Moon colony, or a trip to Mars, we are putting people into true space where radiation damage is a real threat.

The extent of radiation

There are three kinds of radiation to be measured, as reported by NASA:

'Space radiation is made up of three kinds of radiation: particles trapped in the Earth's magnetic field; particles shot into space during solar flares (solar particle events); and galactic cosmic rays, which are high-energy protons and heavy ions from outside our solar system. All of these kinds of space radiation represent ionizing radiation.'

'Flares and Coronal Mass Ejections: When a solar flare or a coronal mass ejection occurs (the two
often occur at the same time, but not always), large amounts of high-energy protons are released,
often in the direction of the Earth. These high-energy protons can easily reach the Earth's poles and high-altitude orbits in less than 30 minutes. Because such events are very difficult to predict, there
is often little time to prepare for their arrival.'

'Galactic Cosmic Rays: These include heavy, high-energy ions of elements that have had all their electrons stripped away as they journeyed through the galaxy at nearly the speed of light. Cosmic rays, which can cause the ionization of atoms as they pass through matter, can pass practically unimpeded through a typical spacecraft or the skin of an astronaut. Galactic cosmic rays are the dominant source of radiation that must be dealt with aboard the International Space Station, as well as on future space missions within our solar system. Because these particles are affected by the Sun's magnetic field, their average intensity is highest during the period of minimum sunspots when the Sun's magnetic field is weakest and less able to deflect them. Also, because cosmic rays are difficult to shield against and occur on each space mission, they are often more hazardous than occasional solar particle events. They are, however, easier to predict than solar particle events.'

'Measuring Radiation: The absorbed dose of radiation is the amount of energy deposited by radiation per unit mass of material. It is measured in units of rad (radiation absorbed dose) or in the international unit of Grays (1 Gray = 1 Gy = 1 Joule of energy per kilogram of material = 100 rad). The mGy (milliGray = 1/1000 Gray) is the unit usually used to measure how much radiation the body
absorbs. However, because different types of radiation deposit energy in unique ways, an equivalent biological dose is used to estimate the effects of different types of radiation. Equivalent dose is measured in milliSieverts (mSv). The mSv, therefore, takes into account not only how much radiation a person receives, but how much damage that particular type of radiation can do – the greater the possibility of damage for the same dose of radiation, the higher the mSv value.'

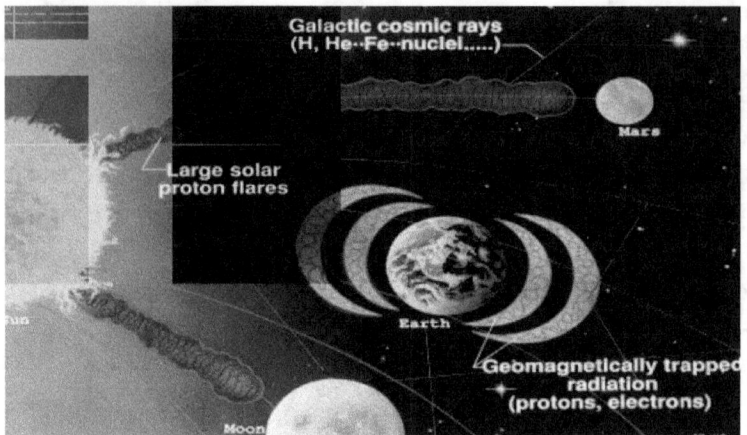

Image credit: NASA

Radiation levels are measured in milliGray's (mGy), and cumulative doses in milliSieverts (mSv).

Radiation varies depending on solar activity, and cosmic ray activity, with typical levels being:

Exposure per 12 months:

Dental X-Ray (per event): 0.01 mSv
Earth surface: 3 mSv to 6.2 mSv
Abdominal CT scan (per event): 9 mSv
Airplane altitude: 20 to 50 mSv
ISS: 160 mSv
Space in the Van Allen Belt: up to 470 mSv
Space halfway to Moon: 400 mSv
Moon surface: 380 mSv
Mars surface: 200 to 400 mSv

These figures can vary; if you are thinking of a trip into space, please take expert advice first!

When you go to the dentists and get an X-Ray on a tooth, what happens? Your dentist and staff set you up, then wear heavy lead aprons and go hide behind a shielded wall. All that just to avoid 0.01 mSv per event. That puts space radiation in context. Space radiation can be 40,000 times stronger than your dental X-Ray!

That is why we need protection from space radiation.

For a human leaving the planet on the space elevator, the highest radiation doses are in the Van Allen Belts, which are massive belts of radiation. While these protect us from general space radiation, the Belts themselves are a risk factor to humans. A rocket has one advantage here: it can move rapidly through the belts, minimizing radiation exposure. But a trip on the space elevator ribbon is a slow one, and humans in a cable car will spend several days each way, within the Belts, so radiation shielding is required for the cable cars.

The problem with rockets is, all methods of radiation protection involve shipping heavy stuff into space, to wrap around a spaceship. Lining a spaceship with Lead is good, though we haven't even mentioned secondary radiation here. Lining the skin with two meters thick of water is also a good idea, but with rockets, they both cost a fortune to get the material into space.

Several strategies are considered for radiation shielding:

* Spacecraft can be constructed out of hydrogen-rich plastics, rather than aluminium.
* Material shielding with hydrogen or water.
* Liquid hydrogen, which would also be brought along as fuel, tends to give relatively good shielding, while producing relatively low levels of secondary radiation. The fuel is positioned in the spaceship skin so as to act as a form of shielding around the crew. However, as fuel is consumed by the craft, the crew's shielding decreases.
* Water, which is necessary to sustain life, can also contribute to shielding. But it too is consumed during the journey unless waste products are utilized.
* Asteroids could serve to provide shielding, in the longer term, as we start to harvest asteroids.
* Magnetic deflection of charged radiation particles and/or electrostatic repulsion is a hypothetical alternative as yet.

Deliveries to space in quantity is the advantage of the space elevator. We can haul things like Lead and water into space, to the Geo Station and beyond, at a less prohibitive price. So our notional plan is for the Geo Station to have a skin made of CNT and a waterproof membrane stretched over metal grids, allowing for a two meter thick, pool of water, to surround the station. Similar methods will apply to everything else including the cable cars.

Because hauling so much water off of Earth is heavy, one solution implies that, at an altitude above Earth, say 1,000 to 2,000 km up, a downdraft car will meet an updraft car and, instead of carrying all that water back down to Earth, the water will get pumped into the skin of the updraft car, keeping the water in space at all times and providing protection when the updraft car emerges from the Van Allen Belts.

NASA has cumulative experience in monitoring the health and exposure of astronauts. As reported in May 2018, it has commenced a longer term study on lifetime exposure:

'Exposure to radiation from the sun and cosmic sources is one of the biggest concerns for astronauts on prolonged space missions, and the Lifetime Heritable Effect of Space Radiation on Mouse embryos Preserved for a long-term in ISS (Embryo Rad) will study mice to examine the possible effects of space radiation on the entire body. Frozen mouse embryos will be exposed to the radiation environment of the International Space Station, and then the embryos will return to Earth to be implanted into surrogate mothers and to live out their lives. Scientists will be able to study any possible changes in the animals' lifespan, cancer development, and gene mutations that may result from exposure to radiation.'

NASA career exposure limits are between 1 Sv (1,000 mSv) to 5 Sv per lifetime. Documentation on these limits are maintained in NASA Space Flight Human-System Standard 3001, Volume 1, Revision A: Crew Health.

'This Standard establishes requirements for providing a healthy and safe environment for crew members and for providing health and medical programs for crew members during all phases of space flight.'

'These requirements apply to all NASA human space flight programs and are not developed for any specific program. However, while some of the existing programs, such as the International Space Station program, meet the intent and purpose of these requirements currently, these requirements may have implications for longer duration missions and missions with architectures and objectives outside of low Earth orbit. Although the requirements are applicable to the in-flight phase of all space missions, it is anticipated that they will be most relevant during long-duration lunar outpost and Mars exploration missions, since the combined ill effects of exposure to the space environment will be of most concern in those mission scenarios.'

The requirements for human space flight health will apply just as much to elevator travel, especially since it will happen in conjunction with an expansion of travel to the Moon, Mars and elsewhere. The positive, as noted above, is that we can haul heavy shielding into space and provide a greater degree of protection to humans.

This will be valuable for space computer systems too. Computers and electrical systems in space are just as vulnerable to being damaged by cosmic rays. In a paper titled Space Radiation Effects on
Electronics by Kenneth A. LaBel at NASA, there is a report of electronic failures following a recent solar flare event:

'In Oct-Nov of this year, a series of X-class (X-45!) solar events took place
– High particle fluxes were noted
– Many spacecraft performed safing maneuvers
– Many systems experienced higher than normal (but correct-

able) data error rates
– Several spacecraft had anomalies causing spacecraft safing
– Increased noise seen in many instruments
– Drag and heating issues noted
– Instrument FAILURES occurred
– Two known spacecraft FAILURES occurred'

(Note: in NASA-speak, "Safing" is a real word.)

3. THE REGIONS OF SPACE

Between Earth and the proposed Geo Station location at GEO are four regions to travel through:

* Our atmosphere, extending to 100 km above sea level. The atmosphere thins out with altitude; it doesn't come to a sudden stop, so even at the altitude of the ISS, 400 km above sea level, there are traces of atmosphere and the velocity of the ISS needs boosting at times to overcome atmospheric drag. Virgin Galactic described their spaceship SpaceShipTwo as reaching space for the first time, though it only flew briefly to an altitude of 82.7 km, which is not "space" by the usual definition. Airplanes fly around an altitude of 10 km.

* The Ionosphere, the layer of the Earth's atmosphere ionized by solar and cosmic radiation. It lies partly within the atmosphere at an altitude of 75 km and extends to 1,000 km above the Earth. It plays an important part in atmospheric electricity and forms the inner edge of the magnetosphere. It has practical importance because, among other functions, it influences radio propagation to distant places on the Earth.

* Van Allen Belts, extending from 400 km to 10,000 km altitude, these highly charged regions shelter the Earth from solar and cosmic radiation, but in the process create a band of highly charged radiation of up to 470 mSv which represents peak radiation for both cable cars and passengers.

* Beyond 10,000 km the cable cars enter the heliosphere, or solar

system space, in which radiation from the Sun is dominant. The radiation dosage varies but is still some 400 mSv in the vicinity of Earth, fading away with distance from the Sun, so it is half that at Mars. The heliosphere is the component of the solar magnetic field which is dragged out from the solar corona by the solar wind flow to fill the Solar System, a three-dimensional form of a Parker spiral that results from the influence of the Sun's rotating magnetic field on the plasma in the interplanetary medium.

Heliosphere

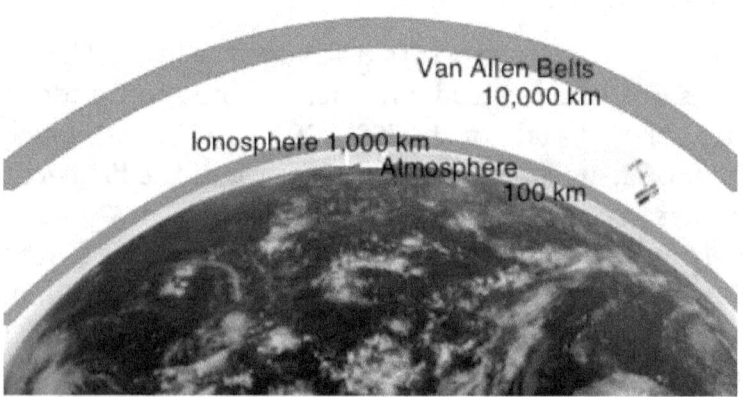

Strictly speaking, true space is about 10,000 km away from Earth, where the Van Allen Belts end and the Heliopause begins.

Although not of practical importance to us yet, we note the heliosphere ends with the heliopause, the region where the solar wind runs into interstellar space, the point recently crossed by our most distant satellites, Voyager 1 and Voyager 2.

These are average values. Sporadic sunspot activity, or spikes in cosmic radiation, can increase this at times, by orders of magnitude, and it is these events which create temporary panic in space. In addition to standard shielding of cable cars and space stations, they may need "panic rooms" which are shielded to a greater degree, providing protection for hours or days, during peak radiation events.

The Ionosphere does not, in itself, present radiation issues for the ribbon, but weapons are being developed which could supercharge the Ionosphere in future. The militaries have been in a race to control the ionosphere for decades.

According to the South China Morning Post in December 2018:

'China and Russia have modified an important layer of the atmosphere above Europe to test a controversial technology for possible military application, according to Chinese scientists involved in the project.'

'A total of five experiments were carried out in June. One, on June 7, caused physical disturbance over an area as large as 126,000 sq km, or about half the size of Britain.'

'The modified zone, looming more than 500km high over Vasilsursk, a small Russian town in eastern Europe, experienced an electric spike with 10 times more negatively charged subatomic particles than surrounding regions.'

'The Sura base in Vasilsursk is believed to be the world's first large-scale facility built for the purpose. Up and running in 1981, it enabled Soviet scientists to manipulate the sky as an instrument for military operations, such as submarine communication.'

'High-energy microwaves can pluck the electromagnetic field in ionosphere like fingers playing a harp. This can produce very low-frequency radio signals that can penetrate the ground or water – sometimes to depths of more than 100 meters in the ocean, which made it a possible communication method for submarines.'

'Changing the ionosphere over enemy territory can also disrupt or cut off their communication with satellites.'

Whilst we can take advantage of shielding in the cable car pods and around the space stations, there will still be electronics receiving full exposure to cosmic rays. There may be parts in the cable cars, and any sensors affixed to the ribbon itself will be exposed. Indeed, the ribbon material may receive sporadic damage, which is one of the reasons for continual monitoring of the entire ribbon system.

4. CABLE CAR DESIGN

Cable car design systems will be developed over the next decade. The moving vehicles have the same systems requirements as any car, train or plane, but with design constraints and opportunities of their own.

The traction system, to provide locomotion, requires a method of holding onto the ribbon while traveling on it, or clamping to it in the case of conveyor belt ribbons. The locomotion system will power large wheels, one of each face of the ribbon, so they can clamp together with sufficient pressure to hold onto the ribbon.

Power systems are likely to benefit from further technological advances. A decade ago, the favored propulsion technology was power beaming, to solar panels attached to cable cars. This may still prove optimal, but a mix of power sources are possible.

While still in the atmosphere, standard gas, petrol, diesel, engines can be used. They draw air from the atmosphere and have the advantage of a high power-to-weight ratio, getting the cable cars up to speed. If we don't use such an engine outside of the atmosphere, which in practice means the first 10 km up, we can envisage a tug-boat separate unit which pushes the cable car up the first 10 km and separates, returning to Earth for the next cable car.

A standard engine could continue to be used beyond the atmosphere, as long as fuel and compressed air tanks are used. This does add to weight, but the first object is to reach 2,650 km above sea level, where the apparent gravity is half that of sea level, expanding the range of power systems possible. Possibly, this is the section where power beaming technology comes into its own.

Solar panel technology will be even further advanced by the 2030s. Already, it is a practical form of power in space. Exactly how low apparent gravity has to be, to get a practical speed, well, that depends on 2030s technology and in-situ testing. But our hunch is, that magic distance of 2,650 km above datum will be where solar panels take over.

At that point, large and light solar panels unfurl to collect power from the Sun. A possible alternative to carrying these up from Earth is a system configuration which allows an outbound cable car to swap with an inbound cable car, passing existing solar panels over. In which case, the solar panels would stay in space, reducing the weight load near Earth. Solar panels in space receive sunlight for approximately 11 to 23 hours per day, the balance being the time the ribbon spends in the night-side shadow of the Earth. This is when power beaming could be used, or battery storage technology.

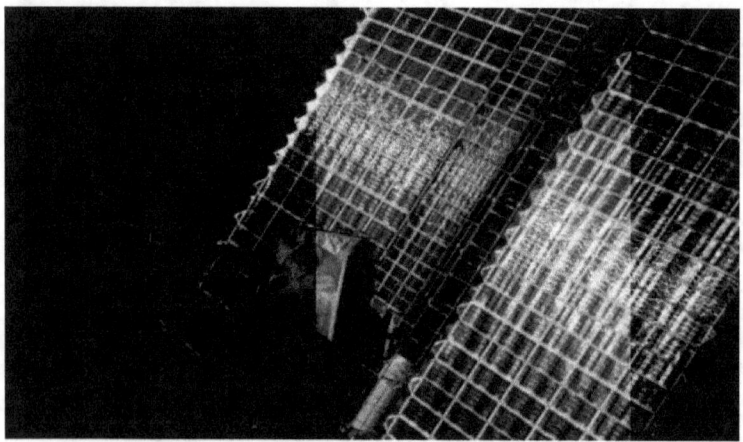

ISS solar panels. Creator: Picasa, Information extracted from IPTC Photo Metadata

Speed? The design speed for the entire journey is around 200 kmh. Way below rocket speeds, but much cheaper. Even so, it's a week's journey to the Geo Station, each way. As apparent gravity reduces, speed can increase, so the cars travel slow near Earth, faster as HS is approached. We won't know an actual design speed until we can do testing in space, but speed is not a critical issue. If a rocket is the bullet train of space, the space elevator is the goods train, slow, but with greater capacity.

A number of cable cars will be on the ribbon at any one time. With a daily service, it would be about 7 outbound plus 7 inbound cars between Earth and the Geo Station (GS), plus more between the GS and the Outer Space Station (OSS). The wheel/locomotion assembly system is designed so cars can pass each other. We describe cars traveling down as the downdraft cars, and cars traveling up as the updraft cars.

Cars can also connect, and travel in tandem. For example, should a car break down, it can receive a tow from the next car.

Cars will be of two types, passenger cars and goods cars. It may be a long time before passengers actually ride the space elevator in quantity. At first the job is to get matériel into space, to the GS and the OSS. Space stations will be assembled at both places, a massive job. The mundane task of shipping air, fuel and water into space may take up half the goods car shipments.

The passenger cars require human-rated cabins, life support systems, additional air, water and waste management, plus docking systems and airlocks for ingress and egress. Add emergency systems, space suits and so on - it becomes apparent the addition of people adds considerably to the mass to be transported.

How many passengers per cable car? At about ten passengers per ton at sea level, we can carry quite a few at a time. What will the passenger experience be? With a journey of a week, we can't just strap passengers in, like coach class on an airplane. They need space (no pun intended) and freedom to move around, plus see outside. The experience needs to be more like, well, a long train trip over the Rockies from coast to coast.

Fortunately, the cable cars can be designed with a massive amount of room. There is a weight and mass constraint, but not a volume constraint. A rocket has a volume constraint since it all needs to fit inside the nose cone, but with the slow travel speed of a cable car, there's no need for tight packing or streamlining.

The passenger car design will likely have three components. In the center, a rigid compartment is attached to the cable car chassis, with all the critical systems, airlocks, fuel, air tanks, computers, pilot controls. Each side of the ribbon, we can have large, even massive, passenger cabins. These will be inflatable, in the manner of the brand space stations. Their shape will be designed around the extended solar panels so as not to shade them.

So imagine two inflatable cabins, each the size of a large house. The design may use Origami approaches to unfold an entire house as the cabin gets pressurized. They will need a "floor" with seats and beds, as gravity still pulls everyone downwards. Anyone expecting instant weightlessness will be disappointed as that won't happen until the GS is reached, though apparent weight will fall off, the further the car travels. Even low gravity is still gravity, so the cabins can have multi-stories with stairs between each, with people climbing up and down, just like home.

Viewing ports will be built in. Everyone will want to see the view, back to Earth (always below the car), into space, and out towards the Geo Station.

With a journey time of a week, separate bedrooms and privacy space will be built in. Wifi connectivity with home will be maintained so everyone can post selfies in space.

The passenger number constraint would be, how many passengers can fit into the center cabin? In an emergency, perhaps a breach with space debris puncturing the inflatable cabins, all passengers would retreat to the center cabin. They would have to stay there, until reaching Earth or the GS, or until they could transfer to the next car.

Another factor is radiation shielding, but that is addressed in the chapter on radiation. Here, sufficient to say that space radiation shielding becomes a severe issue once outside of the Van Allen belt. The Van Allen belt extends from 600 km to 10,000 km above Earth and shelters us from most space radiation. The ISS orbits below the Van Allen belt and is also sheltered by it, which makes living in space look easy. But outside the Van Allen belt? It's like leaving a sheltered cove in a boat and sailing into a storm on the high seas. Radiation shielding is needed.

Can the cars deliver things to LEO? If you are imagining a cable car acting like a space postal service, dropping a parcel off, 400 km above Earth, for the ISS to collect, then the answer is no, at least, not easily. If you think the parcel will just sit there, hanging around in space, for the ISS to come by and collect it, then you haven't understood that the space elevator has no orbital velocity. Drop a parcel off and it will fall back to Earth, since it is not in orbit around the Earth. To drop something off, with the intent of it reaching the ISS, would require a rocket to be attached to the parcel, accelerating it to orbital speed. This could be marginally better than launching the rocket from Earth, but it's complicated.

Add more about power, and using different cars on different sections of the ribbon, so that cars, lower down, can be lighter, and cargo and passengers transfer from one to the other.

The core designs for cable cars envisage a car making the entire trip from Earth to the Geo Station. Since there a number of cars on the ribbon, it includes a design requirement for a cable car to have a mechanism to pass other cable cars. This includes passing, perhaps, a broken-down cable car, or passing a car going in the other direction.

An alternative design is for cars to make portions of the trip only. For example, once past the Van Allen Belts, from 10,000 km to 35,500 km distant, a cable car can be larger and heavier, with large, permanent, solar panels, and plenty of comfort room for passengers.

Between 2,650 km and 10,000 km distant, an intermediate design cable car could be used, with a smaller cable car, using conventional or different power sources, traveling from Earth to 2,650 km away.

The various cars are designed to dock with each other and transfer passengers and goods. The central pod of the updraft car, also used as the emergency and control pod, would dock with the matching pod in the downdraft car to facilitate transfers.

The advantage of this approach lies in cable car design: to minimize weight, cars traveling from Earth can be lightweight and smaller, to reduce the weight burden on the ribbon. Further out in space, the cable car design can take advantage of near-weightlessness to be large and comfortable, and to have large permanent solar panels as power sources.

The cars traveling through the Van Allen Belt, and the cars in the Heliosphere, can have a permanent band of radiation protection, say a two meter deep water skin. This protection is not needed lower down, in the approach to Earth, so it obviates the need to transport all this weight of water, at lower altitudes.

Strain on the ribbon comes mostly at the lower altitudes, the Earth end. Here, anything transported has weight. So any system which reduces weight at lower altitudes is good for the ribbon. So this approach is designed around a minimal design for the lower altitudes cars, while allowing generous parameters for the cable cars that stay in the Heliosphere.

This assumes a single ribbon operation. In Book Two, an alternative design discussed was the Conveyor Belt design, in which the ribbon moves and the cable cars clamp onto it as the means of motion. The advantage of the Conveyor Belt design is: the ribbon movement can be entirely powered by power generation on Earth. But the disadvantage lies in the cable car design: the same cable car must travel the entire trip, so any weight it carries, including heavy radiation shielding, must be carried from Earth.

As CNT material becomes commercially available, it will require testing and computer simulation to determine which factor is the more critical.

5. LOCOMOTION ON THE RIBBON

Now to cover the vehicles that will travel up and down the ribbon. We call them cable cars. If you imagine rotating the cable to the side, by the time the effects of gravity are neutralized, the cars are like driving along a road made out of ribbon.

First stage: building the ribbon

In the long run, we will have passenger cars and goods cars, but initially, with the first ribbon in place, we are focused on pulling up more strands of ribbon using a much smaller vehicle, which we'll call a tow truck. The single R1 cable is 10 cm to 20 cm wide and is already strong enough to bear the weight of light traffic, but we want to build in a redundancy factor which will compensate for minor ribbon damage from meteor or debris impacts. So the first job is for perhaps 9 tow trucks, in turn, to pull up 9 more cables, which will then be stitched together.

Each additional ribbon is attached to the tow truck. At the Earth end, the additional ribbon is on a drum roll which pays out as the ribbon rises. The inside of the tow truck holds fuel, an engine, onboard computer, communications systems and motors to operate two extended arms, one above, one below the truck. These are used if two trucks need to pass each other which they do in a sort of do-si-do dance movement. In addition is what can only be described as a sewing machine, plus a spool of ribbon, plus glue, that will be used by the truck as it descends again, to knit the second ribbon strand to the first one, side by side.

The tow truck climbs up to the Geo Station, where the new ribbon is grabbed by the existing plant there. Then, it descends, commencing the sewing process. The truck travels slowly as it is low powered compared to the later cars that will use the ribbon. It will take about three weeks to ascend, and six to come down again, limited by the sewing process.

A word on propulsion. Way back, the idea was to use a laser, based at the Elevator Space Port, to beam up power to the car, which would have large solar-type panels to receive the power. But possibly a good old fashioned diesel engine with its high power-to-weight ratio, serves the purpose for the first stage, with solar panels in space. So the truck has a tank of fuel, plus a tank of oxygenated air, on board, and the engine is tuned to use the air mix. As the effects of gravity decline, it becomes easier on the engine. Coming back down, light use of the engine is needed, but as gravity takes over, no power is needed for most of the descent.

Once the car reaches about 2,650 km above datum, a change in propulsion is made. Two origami-designed solar panels flex and fold themselves out, into position at either side of the main car, facing downwards. From here on, solar power can be used.

The diesel engine compartment is detachable and has auxiliary wheels, so that it can descend the ribbon independently. It needs almost no power to descend - gravity pulls it down - but it needs power and braking systems to control the rate of descent. During initial tests, the engine compartment will remain with the main car until the power beaming system has been demonstrated to be working nominally. It can act as auxiliary power and can supplement cabin power if required.

Second step: typical usage

The ribbon itself has some extra attachments built in, with, on the edge of the first cable only, small video cameras mounted at one kilometer intervals, capable of being swiveled to view up or down the ribbon. These are connected to tiny solar panels for power and have a wireless data system to beam video and GPS co-ordinates in 3D, down to the Earth Station.

Using the video cameras, and GPS, our computer at the Earth Station can monitor and display the exact position of the entire ribbon in space. If it separates, for example, we would know within seconds. With the cameras, we can also monitor activity on the ribbon, and if a vehicle has problems we can get views from above and below the vehicle.

These will also be used to monitor the locations of the cable cars, and the sensors will enable the cars to keep themselves centered on the ribbon.

For a fixed ribbon design envisaging the car traveling along it, the car is clamped together, on both sides of the ribbon, whether this be by several wheels, or by a caterpillar track. The clamping pressure can be varied, or the wheels can be loosened from the ribbon altogether, for passing maneuvers, or simply to allow the car to coast, out in space, near the Geo Station, where the almost non-existent gravity allows the car to coast along, and the only need is to maintain proximity to the ribbon.

Descending, the downdraft car hardly needs any power, since the gravity pull of Earth allows it to coast downwards. For most of the distance, wheels are used to provide braking, to limit the speed or to come to a stop when docking with, or passing, other cars.

That business of "passing" needs further explanation. On Earth, on a train line, two trains can pass each other. Using points, one train moves onto another parallel train track and they pass each other.

It's not the same in space, on the ribbon. Trains are kept connected to the train line by gravity. In space, a cable car doesn't drive "on" a ribbon. It requires clamping to the ribbon, to maintain proximity. So, it isn't as simple as the updraft car traveling up one side of the ribbon, while the downdraft car travels down the other.

A number of passing concepts have been put forward, but the main concept is easier to understand if you imagine a number of people on a wooden ladder. To pass, one person could climb over the back of another, and the ribbon concept is similar.

In addition to the core wheels or caterpillar track, each cable car also has two extended arms, one above, one below, or if you prefer, one in front, one behind. The extended arms also have wheels, and clamp on the ribbon, permitting the main wheels to be unclamped. The system is designed in an origami fashion, to enable the cars to inch past each other, until they can engage the main clamps again.

A variation on this, is for the core central pod to have a firm track over its back, or ceiling. This allows for the main wheel clamp to widen, and ride over the cabin of the other car.

If, instead, the respective cars are docking, then no passing maneuver is required. Instead, it is a docking maneuver which calls for the cars to line up and dock.

6. EMERGENCY RIBBON MANAGEMENT

Moving ahead to when the entire cable has been stitched together, we will have a flat ribbon of several meters' width. Now, we don't want to lose that cable due to separation factors.

We've mentioned ways of reeling in a separated cable. But on completion, we can obviate the need for that by installing an emergency cable, a meter or less in width, capable of carrying traffic. This is a secondary ribbon, parallel to the main ribbon, but not in tension, loosely attached to a cable car, providing an emergency ribbon to travel on.

Picture this. The main cable is in place. It is, in round figures, 35,500 km in length, and is in tension, that is, taut, like a guitar string. The emergency cable will be 37,000 km in length, so it will be slack. If it is hit by a large object, it is capable of more lateral movement and less prone to breakage. Cars traveling on the elevator will be connected to both, though, in normal circumstances, they will not actually be attached to the emergency ribbon, so much as loosely holding it. Think of a person climbing a rock-face, with an emergency rope threading through a carabiner. In the event of a main ribbon separation or any other relevant issue, the vehicle can clamp firmly onto the emergency ribbon and continue a journey on that.

In addition, we will have procedures for repair vehicles, based at Earth Station and Geo Station, to travel along the emergency ribbon, pulling the main ribbon back together, and a mechanism for repairing the main ribbon.

As a result, the level of safety will be vastly improved, and the risk of losing the elevator system reduced, to the point where the elevator will never be lost.

What's the most likely cause of a cable separation?

Two classes of issues:

First, actual breakage would most likely be caused by a collision. It will repeatedly be hit by micro-meteorites, the size of a grain of sand, but traveling fast. These in themselves would not cause a breakage.

Larger objects such as a large meteor, asteroid, space debris or space station are another issue. As objects get larger and faster, the risk of separation increases, which is why we will make live use of CNEOS and Satview to track all known objects, and why we can move our ship around to deploy the ribbon away from risk incidents.

Second would be deterioration in the ribbon itself but as yet, of course, we have no information on CNT longevity in space. That is yet to be studied in situ.

We drop the ribbon down from space. This is the opposite of fly fishing: we are the fish, and we have to catch the hook when it comes down.

Because of gravity, the ribbon, R1, which will only be 10 cm to 20 cm wide and almost nothing thick, will come down at the Equator, in open ocean 3,000 km south of Hawaii, assuming this to be the target location.

It will have a light grapple hook and magnet on the end, plus sensors that will give us a continual readout of the position via GPS. This will enable the ship to be in the correct position as it lowers to Earth.

Somewhere in the last few hundred meters, a helicopter with a long probe designed to couple with the ribbon, would be good for initial capture, and it can then be guided down to the ship. Once there, it is quickly attached to a large drum roll and it starts to be wound in. Once we have about 100 km or so wound onto the drum, we instruct the Geo Station to cease paying the ribbon out, and the ship sets sail towards the designated Elevator Space Port location, out in the open ocean.

Why wind so much ribbon around the drum on the ship?

We are towing the ribbon away from the equator, so the ribbon needs to lengthen if the Geo Station is to remain in place. Much of that will be catered for by paying out more ribbon from the Geo Station beyond GEO, but we want the flexibility to do the same from the Earth Station.

Initially the ship needs little more than the drum roll and engines to roll the ribbon up or down, plus lots of room for stores, and for the first goods cars that will be attached to the cable. Later, we will expand facilities on board, passenger facilities and so on. At this point we have reached our mission target, demonstrating that the CNT ribbon will, in fact, support its own weight.

Next, a few comments about management issues at the Earth end, and procedures for disconnecting and reconnecting the ribbon from and to the Elevator Space Port. The thing we most want to determine is the resilience of the CNT ribbon to weather conditions. Dr. Edwards research shows that, although the material is strong, it could be melted by a lightning strike, and a lot of work went into identifying places on Earth that never have thunderstorms, that never have typhoons or wild weather.

Analysis of NOAA weather data showed that there were eight locations globally that are calm islands of weather, where a sailor could be marooned forever. A number of them were discounted for political or security reasons, leaving a few preferred ocean locations. One is the Pacific Ocean south of the Equator, stretching west from the Galapagos Islands towards, but stopping short of, Fiji and New Zealand, where the good weather extends west across most of the Pacific, while another is west of Australia in the Indian Ocean. Ideally we'd prefer a location in the Pacific Ocean just west of the US, that could be serviced from LA: there is one possible location for this, but the weather is less predictable as compared to the southern Pacific Ocean and the Indian Ocean.

Recovery procedures

Possible recovery procedures in case of ribbon breakage, for the passenger cars, depend on the car altitude.

Within the atmosphere, a solution is to detach the car from the ribbon and deploy parachute(s) to drift back to the ocean where the car can float.

For a breakage when the car is above the atmosphere, anywhere on the ribbon, but below the ribbon breakage, it jettisons mass in the form of goods payloads, and transfers the load to the emergency ribbon, referred to earlier, on which it can descend.

For a breakage when the car is above the atmosphere and above the ribbon breakage, it continues the journey up to Geo Station. With the ribbon loose, this will be a wilder journey, as the ribbon won't be taut, but it will have the emergency ribbon to assist.

Assuming it can arrive at Geo Station, and assuming that previous stockpiles of air and emergency supplies can be accessed there, the recovery then becomes a matter of (a) a rocket spaceship launch and recovery procedure or (b) a launch with a replacement ribbon, awaiting the entire procedure of putting a new ribbon in place, or (c) a transfer to another ribbon, assuming two or more Space Elevators have been erected.

To obviate the risks for a stranded passenger car, ideally, we will have transferred an entire replacement ribbon to Geo Station, which is available for deployment in such an event.

7. POWER DELIVERY

The climbing cable car vehicle needs a power supply. This needs to supply megawatts of power to the drive system, equivalent to the power needs of a road truck.

Where can you get such power in space?

A decade or two ago, delivering power appeared to be a significant problem and a lot of work has gone into futuristic methods of beaming power to a cable car, such as a laser focused on solar panels mounted on the car.

The beaming of power may be practical, one day in the future, but we are back in the realms of scifi with this one. Advances have been made, but not to the extent needed for a space elevator.

Current state of the art appears to be a plan by the US Army to beam power to airborne drones, as reported by nextbigfuture. The announcement was made in late 2018 to launch a system in 2019:

'By early 2019, the US Army will start providing power to a drone via a laser beam. They hope to be powering airborne drones by 2020 as long as the necessary regulatory process is approved.'

'DARPA will try to provide power to drones at a range of 500 meters. DARPA will demonstrate the feasibility of recharging the batteries on board, in flight, by using a laser light source, allowing for indefinitely long flight times by using concatenated "Fly" and "Fly & Charge" cycles removing the need to land to refuel.'

Powering something 500 meters away is far removed from powering an elevator cable car 30,000 km away, or even above the atmosphere at 100 km away. In fact, the web page displays a graph showing the current state of the art:

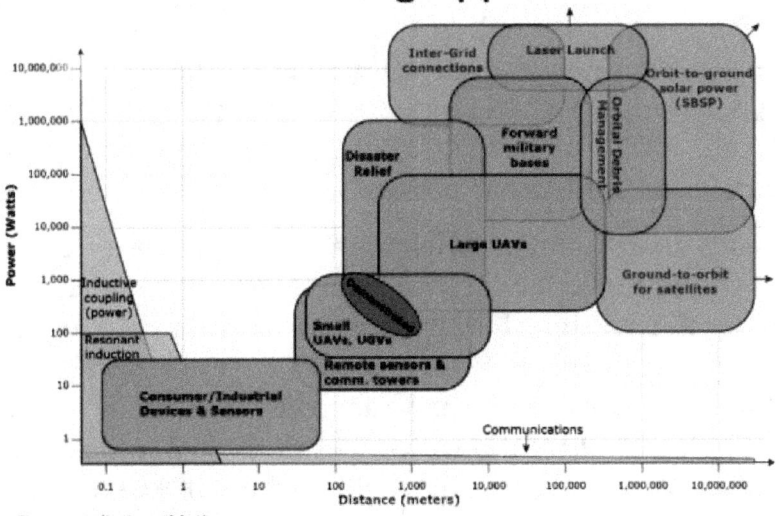

Figure credit: LaserMotive

Image: State of the Art, 2018

In 2018, demonstrated applications have stretched only to delivering around 500 watts at 500 meters. Space elevator requirements are more demanding, in the box at the right of the graph marked "Orbit to ground solar power", and one bar further along the logarithmic X axis. Whilst acknowledging the fast pace of development in the electric/battery/generation industries, we are yet a long way from this concept. Perhaps by the 2030s the industry will have advanced, but meanwhile it cannot be utilized.

It shouldn't need stating, but the obvious isn't always obvious: we can't attach rocket engines to the cable cars, as the heated rocket exhaust would be emitted adjacent to the ribbon, destroying it.

Based on present technology, which is still advanced on that of a decade ago, the best approach may be a mix of propulsion techniques. As previously noted, variations in power load can be assisted if different car designs are used at different altitudes, minimizing the weight load near Earth, and maximizing solar panel generation way out in space.

The power load is highest, closest to Earth, due to gravity. Power needs decline as cars ascend, with half power being sufficient 15,000 km up, declining to minimal power by the time GEO is reached. Coming down, power needs are low as there is a gravity assist. Indeed, within some 15,000 km of Earth, the need is braking to arrest speed, rather than power.

Noted already is the time involved: the ascent will take a week or more. It is the equivalent of driving uphill, a journey which may start off slow, but accelerating later, with an average speed of 200 km/hour postulated. For the return journey, it's the equivalent of driving downhill. What speed can be achieved? No idea as yet, except it will be faster. Currently we assume an average speed of 300 kph for the return journey, which would be about five days.

In theory, a journey that free-floats to Earth, falling, can be postulated, which could cut the journey time to less than a day. But the constraints on speed need testing. In particular we flag two problems. One: the car wheels will be kept loosely attached to the ribbon; will friction and heat generation be a problem? Two: when a car, descending at several thousand kph, needs to slow down, the wheels will need to be clamped onto the ribbon. Besides the issues of friction and heat, the sudden addition of mass to the ribbon may cause it to stretch. We don't yet know at what point the mass addition becomes a threat to the ribbon integrity.

It can be imagined a parachute is added to the returned car to slow it down, but the proximity of the ribbon introduces additional risk factors: friction and tangling.

So, the power load needs designing for peak use, which will be on commencement of ascent. Let's state the issue here: an elevator car requires significant power, especially on commencement, but we have weight constraints. Engines can be heavy, and also require fuel tanks. It's the fuel which tends to add the most weight.

Unless some significant advance in power beaming is made in the next decade, a mixture of propulsion sources may be used, as follows:

At launch, the car is at sea level (ASL), and will be traveling slowly, from zero to perhaps 50 km/hour in the first minutes. A car engine, in the first 500 meters say, could be fueled by gas, or electricity, by a connecting gas or power line from a generator which stays at sea level in the terminal. The line can be auto-disconnected at 500 meters or whatever height works.

At that point, a petrol/gas engine can take over. This could be in the car itself, if the weight load is not too heavy. Or, below the car, we can position a shunting engine, which has the engine and fuel tanks. The key thing is, such an engine uses oxygen from the atmosphere, and any normal engine can propel the car, taking it up to about 75 km ASL. If a shunting engine is used, it would have air tanks as well as fuel tanks and could continue propelling the car, as long as air and fuel hold out.

Think how far you drive on the road with a tank of gas - designing a shunting engine to travel 500 km to 1,000 km can be done using existing technology. Once fuel is exhausted, it can be designed to crawl back down the ribbon, to the Elevator Space Port at sea level.

At 1,000 km or so ASL, we are way past the ISS orbit. There is no atmosphere to use, so we switch power sources. At the surface of the Earth, the acceleration due to gravity is about 9.8 m/s2. At 1,000 km above Earth's surface it is 7.3 m/s2, about 75% of what it is ASL. So loading is still significant. At this point, the most likely option is an onboard nuclear powered engine, based on designs used in submarines.

Apparent gravity disappears rapidly with altitude. (Take your bathroom scales with you for instant weight-loss!)

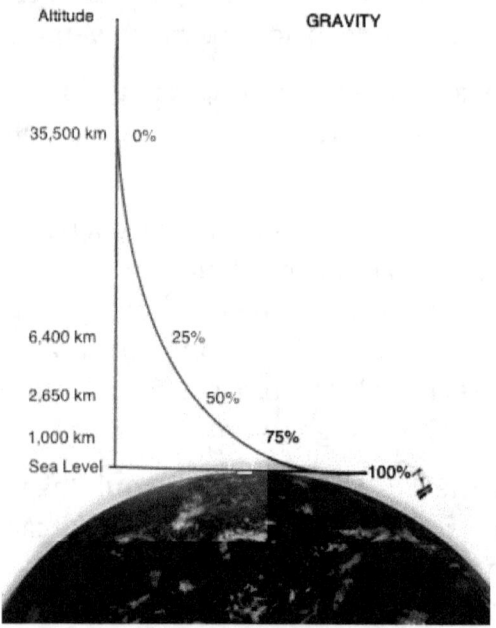

As the car climbs the ribbon, the power load reduces - dramatically. That same curve in the graph can be thought of as the weight that the cable car is carrying, reducing with altitude.

Once the car is about 2,650 km ASL, gravity has already dropped to half what it was at ground level. The car will have traveled slowly. It may take a day or two to reach this seemingly low altitude, although even this is above the range of low earth orbits (LEO), where the ISS orbits. As gravity reduces, we can take advantage of being clear of the atmosphere and start to deploy solar panels, to power the car by solar power. This will take the car all the way to the Geo Station at GEO. From 2,650 km altitude, the car still has about 33,000 km to go, but as gravity falls away, it can travel faster and faster, so about another five days to finish the journey.

For space elevator purposes, we can treat altitudes above 6,400 km as being in space, since the payload weight has dropped 75%. The atmosphere is a long way behind us, and we can deploy large solar panels for power. So, most power management issues relate to the 6,400 km closest to Earth, where we need boosted power to get out of the gravity well.

Note: in the above, we describe the altitude of the car, not the orbit. The car and the ribbon are not in orbit around the Earth - from the surface of the Earth, it looks like the ribbon stays in one place relative to the observer, whereas the ISS seems to whizz past in minutes. The ISS stays up there by orbiting the Earth, whereas the ribbon stays in place, pulled by the centrifugal force on the upper ribbon.

As noted earlier, volume carried is no problem; it is weight, or specifically, mass, that is the constraint. So, we will be able to deploy very large solar panels to generate electricity. As the ribbon rotates along with the Earth, it is in direct sunlight for half or more of the time. Here, we don't speculate on the size or specification of solar panels used, since the technology is developing at a fast pace. Compared to when our first book was written 12 years ago, solar panels have halved in price and weight, while doubling their conversion efficiency. At the time of writing, 2018, we envisage another decade, at least, to pass, before constructing the space elevator, and solar energy technology will likely have advanced further still. By the time we get to launch a space elevator, solar panel technology may be the best source of energy.

Solar panels are now used widely in the satellite industry. In December 2018, SpaceX hosted a public viewing of Crew Dragon's integrated solar panels for the first time. The Crew Dragon is a crewed launch vehicle that is capable of ferrying astronauts to and from the International Space Station.

Image: SpaceX "Crew Dragon" solar panels

On display were the spacecraft's built-in solar arrays, shown for the first time. Reports Mike Wehner:

'As you can see, the solar panels are formed to match the shape of the ship itself, meaning that the spacecraft won't have to deploy and external solar array to gather sunlight. These panels are covered by a shroud that is jettisoned, exposing them to sunlight after the ship is already outside of Earth's atmosphere.'

'It's an interesting solution but actually getting the solar panels to fit while accounting for the fact that the array will expand and shrink depending whether it's being hit with direct sunlight caused plenty of grief. The end result, however, is undeniably eye-catching and if it works as planned it'll be a major win for SpaceX designers.'

Of course, less than half the solar panels will be in sunlight at any one time, zero to 50% of them, depending on the vehicle orientation. But it is a good example of where we are going with solar power.

Solar panels are becoming lighter and more flexible in design, able to collect more power. By the time we get to the 2030s, around the launch window of the space elevator, we will likely be able to collect a large proportion of required power, if not all, from solar panels. These can be unfurled on both sides of the cable car, using an origami based design to bend and adjust the panels to maximize their solar exposure.

As the Earth turns, it experiences day and night, and the space elevator ribbon will experience day and night too, so solar power collection is not a 24 hour thing. But, unlike on Earth, peak solar capture happens as soon as the ribbon leaves the Earths' shadow and continues until it re-enters the shadow. For cable cars, the further from Earth they are, the less time they spend in Earths' shadow.

With battery storage systems likely to have matured also, a combination of solar power and battery storage would be the most desirable option, if it can meet the needs of a cable car. To the extent it doesn't, we introduce supplementary power sources, whether they be fuel-burning engines, beamed power, nuclear power or other sources.

8. THE ELEVATOR SPACE PORT ON EARTH

This is the JFK or Heathrow for space travelers, Cape Canaveral for coach or economy class.

Where will it be?

If you think Cape Canaveral is the only rocket launch site in the world, think again. There are hundreds of launch sites that have been used, are still in use or are proposed.

Of these, the most significant rocket launch facilities are:

* Cape Canaveral Air Force Station, FL, USA
* White Sands Missile Range, NM, USA
* Baikonur Cosmodrome, Tyuratam, Kazakhstan (previously USSR, Russian operated)
* Plesetsk Cosmodrome, Russia
* Wenchang Satellite Launch Center, Hainan Island, China
* Tanegashima Space Center, Tanegashima Island, Japan
* Satish Dhawan Space Centre (Sriharikota), Andhra Pradesh, India
* Guiana Space Centre, Kourou, French Guiana (French operated)

Eight major rocket facilities, but they are all on land. Since our space elevator needs to be mobile, it will operate from a base on the open ocean.

The Earth-end of the ribbon will need attaching to something as an anchor, not to hold it up, since the ribbon hangs down from space, but to stop it floating away and to provide docking and boarding facilities to get on and off of the ribbon.

The anchor station will initially be little more than that: a floating platform, to which the ribbon is connected, and which is heavy enough to dampen down oscillations in the ribbon, and keep it connected to Earth. As noted above, it will be a large ship, enabling the station to be positioned at a desired point in the ocean. At that point it will be connected up to a whole network of floating platforms that together create a huge docking facility: the "Elevator Space Port".

One of the major considerations that drive the anchor is required location. An ocean based location is preferred, since this allows for movement of the terminal, perhaps to drag the ribbon out of the way of an asteroid or space-debris. It will be necessary to move the ribbon from time to time, and ribbon oscillations may create a need for movement, so a land-based station is considered impractical. Also, the calmest locations on Earth, with few or no storms, hurricanes or lightning strikes, are out at sea, generally south of the equator, in the Pacific and Indian oceans.

The Elevator Space Port will be a floating location on the ocean. Humans are getting pretty good at building large ocean-going vessels and platforms.

The design of floating oil rigs is of particular interest to space elevator terminal designers, since the engineering requirements are similar. Coping with extreme weather conditions, these rigs demonstrate our ability to cope with major challenges.

The largest floating oil rig in the world is the Russian Berkut rig.

'This 200,000 ton rig is situated off the Russian Pacific coast. While other oil rigs are larger beneath the surface, Berkut boasts the biggest upper part in the world. Translated as 'Golden Eagle',

Berkut is built to withstand subarctic conditions: including 60 foot waves, and temperatures of up to minus 44 degrees celsius.'

Berkut Oil Rig. Image source: EnglishRussia

The proposed Elevator Space Port may not be as big as Berkut, at first, but the design of Berkut provides useful data.

The method of containing the Elevator Space Port in a given position is also under research. When it has to be moved, it can be connected to tugs, though tugs much larger than used in ports, the size of modern oil tankers, with powerful engines.

When the terminal is to remain stationary, we learn from the methods used to tether oil rigs. There are ten main methods used to tether oil rigs to a location.

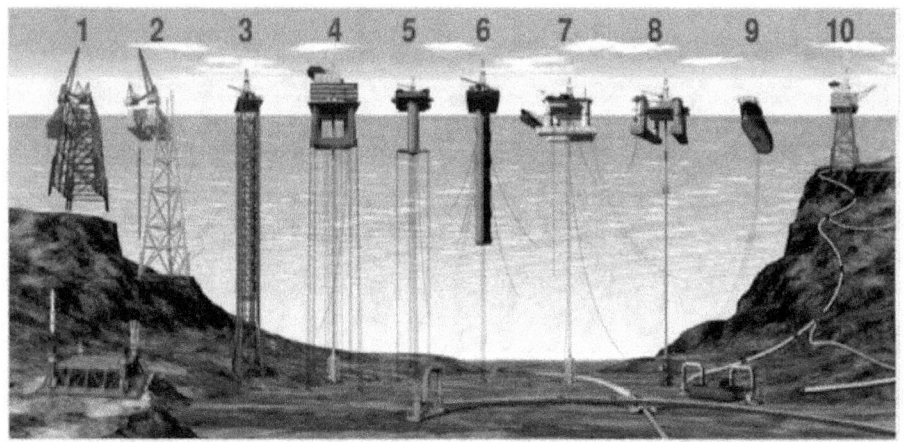

1, 2) conventional fixed platforms; 3) compliant tower; 4, 5) vertically moored tension leg and mini-tension leg platform; 6) spar; 7, 8) semi-submersibles; 9) floating production, storage, and offloading facility; 10) subsea completion and tie-back to host facility.[16]

Oil rig platform tethering methods. Source Wikipedia

Of particular interest for space elevator terminals is method 4, a vertically moored tension leg platform. Wikipedia reports:

'TLPs are floating platforms tethered to the seabed in a manner that eliminates most vertical movement of the structure. TLPs are used in water depths up to about 2,000 meters (6,600 feet). The "conventional" TLP is a 4-column design which looks similar to a semisubmersible. Proprietary versions include the Seastar and MOSES mini TLPs; they are relatively low cost, used in water depths between 180 and 1,300 meters (590 and 4,270 ft). Mini TLPs can also be used as utility, satellite or early production platforms for larger deepwater discoveries.'

9. ELEVATOR SPACE PORT LOCATION

Next are details of the Earth Station location. This is covered in detail in Book Four of this series, but with a brief overview below.

Lateral movement may be needed to avoid debris, asteroids, or the ISS or anything else in orbit. To move the ribbon out of the way of the ISS, for example, needs only for the ship to move a few miles/km north or south, though for safety purposes we'd take it a hundred km or more out of the way.

We do not anticipate that the ribbon will break. As time goes by, we will be reeling more and more ribbon up, adding to the existing ribbon, so it will get stronger. However, we will have an MPC, a Mishap Preparedness and Contingency Plan which we are continuing to develop in accordance with NPR 8621.1, which basically says this: In the event of ribbon breakage, the Earth Station will need to be able to reel the remaining ribbon in, at a speed greater than the ribbon will be falling to Earth, to prevent it wrapping around the Earth to the west, due to Earths' rotation.

That is easily managed, while to the west of the assumed target zone, (if it is in the Pacific) is empty Pacific Ocean and some small islands, until you reach Australia. The remaining portion of the ribbon above the break line will still be attached to the Geo Station, at GEO, and that will have the capability to wind up the ribbon balance to GEO and store it on a drum.

In this event, the Geo Station, and the Elevator Space Port on

Earth, will still be connected via the emergency ribbon, and this can be used to assist in deploying a replacement ribbon.

The USA has a launch facility at Cape Canaveral. Why not use that?

It's in Florida, on the east coast. The east coast of every continent is the coast that gets the worst weather, typhoons, storms, lightning, floods. We don't want to base the Elevator Space Port on land, we want the flexibility of movement that a ship gives us. Plus, the weather at Florida is bad, on a regular basis.

Of course, the USA launches rockets from there with no problem.

But a rocket launch only requires a ten minute window of calm weather to launch. How many times have countdowns been stopped because the wind speed picked up at the last minute? But for the space elevator, we're not talking a ten minute launch window, we're talking a 24/7 operation that must be able to operate without interruption.

As far as we are concerned, Cape Canaveral is about the worst place to put a launch facility. It was only built there in the first place as it was the most southerly part of mainland US, and a senator wanted it in his patch.

Lightning over Cape Canaveral. Image source Pinterest

Accu weather reports: 'Cape Canaveral was not the first choice by the United States when they began launching rockets, and the weather may have been an influence in that decision. The weather does not always cooperate for rocket launches as more thunderstorms erupt in Florida per year than in any other state.'

'Lightning from thunderstorms can be catastrophic to a rocket if it strikes in mid-flight. In 1987, lightning struck the AC-67 rocket less than a minute after liftoff, causing it to explode.'

Do objects fall to the east or west of a launch site?

The one advantage of Cape Canaveral is for rockets. The Earth spins towards the East - think of the sun rising in the morning. Really, it is not rising, it is the Earth rotating towards it. Rockets are launched with a trajectory towards the East, to add the rotational velocity of the Earth to its speed.

If such a rocket fails, then it will also fall towards the East, in this case, away from populated areas and in the Atlantic Ocean. But with the space elevator, we have the opposite issue. It is not moving east, it is, apparently, stationary above a point and if it failed, the Earth would rotate away from it, making it seem as though the elevator ribbon would fall to the west. Hence, for the space elevator, it is an advantage to have clear space to the west.

Initially, our shortlist of locations was based on the lightning risk. However, with the newer CNT and further planning, the risk has lessened. Even so, we don't want a massive electrical charge threatening the ribbon. So, we have constructed MPC mitigation scenarios as follows:

Electrical hazard mitigation options:

1. Earth locations with little or no electrical or lightning risk
2. Being able to disconnect the ribbon from the Elevator Space Port and winch it up to 12 km above sea level, above the atmospheric storms, then to lower it again on clearance.
3. Coating the ribbon with an insulator or using a different material for the lowest 12 km.

Note 1: a different material for first 12 km is not yet feasible, but once sufficient ribbon has been uplifted, it could be.
Note 2: Placing an insulator in the ribbon, 12 km up. We are still evaluating potential materials. The insulator needs to be strong enough to support all the weight below it.

Moving back to the capture of the ribbon, and operating south of Hawaii, there are some legal issues there.

Just to summarize, the ribbon will descend to Earth at the Equator, no matter where the intended final destination is. That means it will would likely descend into international waters. We will need to declare an exclusion zone and will need to move significant US Navy ships there, to protect it.

There is legal precedent. This procedure has been used in the past when spaceships have splashed down in the ocean, but there is a risk, given the high stakes, of other governments trying to sail, fly or use drones nearby for spying or offensive purposes.

Plus, if the Elevator Space Port is to be in international waters, all the same problems apply, only they are magnified as it is intended to be a permanent position which will be subject to legal challenge. These are all manageable. So, a location that can be defended from legal and military challenges is required.

A location west of Perth, Western Australia, has been suggested as one candidate.

Alternative locations, with logistical advantages for the USA, would be Hawaii, north of the equator, or, south of the equator, in the Pacific Ocean, serviced from Howland Island, Jarvis Island, Kwajalein or a number of islands and atolls around there, including the temptingly sounding Starbuck Island.

Other locations include three in the Atlantic Ocean, one off of French Guiana, one near Ascension Island and one near the Cabo Verde islands. See Book Four for more details.

10. THE GEO STATION AT GEO

Some details about the Geo Station: Initially, it will be a basic set of hardware and software, most important being a lightweight crane beam and grab. As there is no gravity effect there, it's experiences weightlessness, it is not weight that is the issue, but mass, however we don't have to rush operations. The crane is used for removing deliveries and placing them in position, while AI computer controlled assembly vehicles and cranes will be doing the initial construction work.

As previously mentioned, a key operation will be to grab new ribbons and add them to the clamping systems that will hold all of the ribbon at the Geo Station. A passive aspect is mass. As far as we are concerned, the more mass the Geo Station has, the better. It becomes more resistant to the mass of the cable cars as they move.

We've talked about the Geo Station being at GEO, but this is actually a fulcrum point. In the long term, we'll add a counterweight, much further out, so that the Geo Station can remain at GEO. In the short term, we'll move it some way further out than GEO, depending on the mass load on the ribbon. When we are launching a car or tow truck, Geo Station will be positioned about a further 5,000 km away and it will reel in to GEO, matching the balancing mass of the car or truck as it moves up the ribbon. Later, we'll position more sophisticated control methods, but this is how we'll manage the initial launch and construction phases.

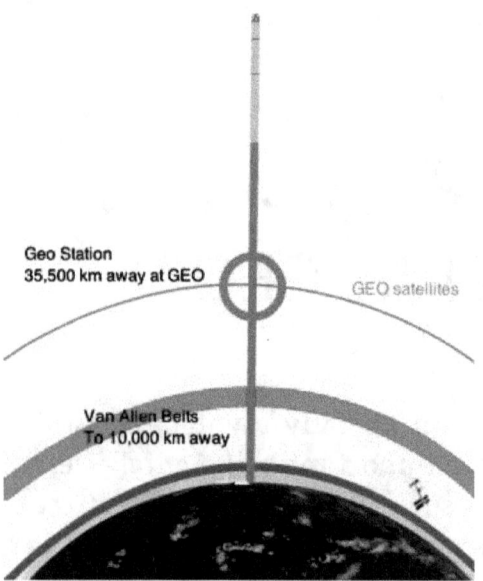

Geo Station
35,500 km away at GEO

GEO satellites

Van Allen Belts
To 10,000 km away

What is the purpose of the Geo Station?

The Geo Station (GS) is the center of the enterprise. It is situated at GEO, in a stable position fixed above Earth, and is the point where apparent gravity is zero and weightlessness is experienced. This is the destination for cable cars inbound from Earth and is the departure lounge for cable cars outbound to the OSS, the Outer Space Port. It will be large, perhaps kilometers across, by the time it is completed. Brightly lit, it will be like another start in the sky, when viewed from Earth.

It will be our first true space station, dwarfing the little ISS, which is barely in space, 400 km from Earth. 35,500 km away, well past the Van Allen Belts which only extend 10,000 km from Earth, the Geo Station is really in space, in the Heliosphere.

This is the pivot point on the ribbon system, where the gravitational attraction of Earth is offset by the centrifugal force of the OSS and its ribbon. It is the one stable point in the ribbon system. If the ribbon were to break, one side could fall to Earth while the other side could travel out into space, but the Geo Station would be capable of staying exactly where it is, even without connecting ribbons.

The stability of the ribbon system is aided by the stable mass of the GS. The more mass at GS the better as its inertia will act as a semi-stable fixed point for the ribbon, keeping the ribbon in tension, as near a straight line as is possible, all things considered.

It is at GEO, which makes it a jump-off point to launch GEO satellites, retrieve or service them.

It can be a place for spaceships to arrive at or depart from. The OSS will serve this function too, but the orbital velocity of the OSS is different to the GS. Depending on where a spaceship is going, one or the other may have advantages. For trips to and from the Moon, for example, the GS has a lower orbital velocity relative to the Moon, so GS/Moon transfers will be easier. However, the OSS can add velocity, in the direction of the orbit of the Earth, giving a boost to interplanetary spaceships.

Because it is located at a stable point, with zero apparent gravity, the GS can be as large as we can build it. The bigger the better; we want lots of mass here to help with stability. Eventually, this can become so large as to be kilometers or more across. It will be the sort of space station featured in space fiction movies, a Star Trek kind of thing.

One advantage of a large GS is dealing with space radiation. If it grows as a sphere, square or rectangle, then surface area shrinks relative to volume. The surface will support an immense amount of radiation shielding, perhaps of water and concrete, while the inside volume will come to resemble a zero gravity city, with areas of offices, warehouses, hospitals, waste management, residential space and community facilities. Food may be grown, perhaps with small farms and community parks.

Heat management is also an issue. Space is cold and the GS will radiate heat away, so passive and active heat shielding measures will be used.

The GS can be constructed with rigid steel girder frames. The external walls can be covered with CNT material stretched around it, in two layers, filled with water as a radiation shield, lined with foil to help retain heat. Constructing the frames will be activity somewhere between assembling Ikea ® furniture and doing Origami. The steel girders are used for strength, but they don't have to be big and heavy, just strong enough in compression to maintain the load.

A number of power sources can be considered. A nuclear power plant could be used, or even conventional fuel engines, using fuel lifted from Earth. But, as a stable space station, solar panels will be an obvious choice, extended as big as we need. Many a science fiction story has been written about solar panel engineers in space, drifting around, fixing these things, though in practice an AI based computer system will be managing most of it.

It will have massive tanks storing air, fuel, water and specific gases like oxygen. Air and water tanks will be inside the walls, shielded from radiation, but fuel tanks could be outside.

Besides these are terminals and docking ports for cable cars, as well as external spaceships. Transfer of goods and people, maintenance and fueling, will be routine activities.

11. LIVING IN THE GEO STATION

Life in the ISS gives an idea of what life will be like in the Geo Station. The biggest difference is that, whereas the ISS is small and cramped, the Geo Station will be large and roomy, resembling a zero-gravity city.

Viewing ports will let people see into space, towards the Moon at the right time of day, and back to Earth, a long way below. Viewing ports don't just act as a window; they form a gap in the radiation shielding, so a number of windows would be positioned at optimal viewing points, but time spent looking out of the windows would be limited, to minimize radiation exposure.

Because the Geo Station is so large, and surrounded by protective water-lined walls, most workers would spend their shifts well inside and, after having gotten over the novelty of looking out of the windows, would work, play and sleep indoors. Long term workers might almost forget they are in space, were it not for the zero-gravity. After the novelty of being in space wears off, day to day life won't be much different to life on Earth.

Eventually, tourists will come and go, but the initial crew will be workers. Inside, it will grow from a small and cramped workspace like the ISS, into a large 3-D city over time. Initially there will be two crews, one crew managing arrivals, departures and maintenance of cable cars and the ribbon, another crew dedicated to construction work to expand the Geo Station.

Being weightless, moving around will be a different experience. You can launch yourself in any direction and drift there. However, we don't want collisions, so there will be guide ropes made of CNT. Hold on, just like abseiling, but loose enough to travel.

Perhaps new zero gravity games can be imagined.

It would be necessary, at some time, to build a ring in part of the GS, a large one that rotates, generating centrifugal force to provide the illusion of gravity. The impact of no or low gravity on the human body is well documented, and this would be a way of generating a pseudo gravity for humans. This ring could be as large as can fit in the outside walls. As humans work their way to the edge of the ring, it will feel like a rotating floor, where they can exercise in full gravity. But as they leave the floor of the ring and approach the hub or center, the pseudo gravity fades away and weightlessness returns. However, it is done, something like this will be needed.

12. THE OUTER SPACE STATION

Exciting though it is, the space elevator doesn't end at the Geo Station. The ribbon extends beyond GEO, to about 100,000 km from Earth, 64,500 km from the GS, until it reaches another space station, right at the end of the ribbon. We nicknamed it the FarOut station, but that name was taken, albeit dropped again, in December 2018, for a newly discovered dwarf planet.

So, Outer Space Station, it is, because this station is the connection for humans to outer space. After all, the space elevator isn't an end destination, it's a process, the means to leave the planet. The destinations are out there: the Moon, Mars, everywhere else in the solar system and beyond.

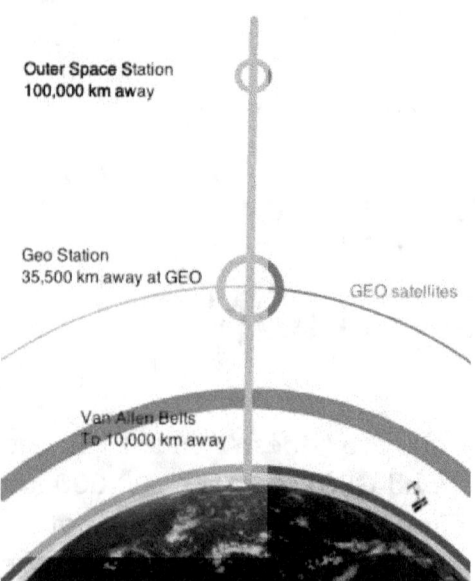

Outer Space Station
100,000 km away

Geo Station
35,500 km away at GEO

GEO satellites

Van Allen Belts
To 10,000 km away

Building the OSS

This will be our second space station, a quarter of the way to the Moon.

Out there, the orbital velocity of the ribbon creates a centrifugal force pulling away from Earth. This is how the space elevator ribbon stays in place. One purpose of the OSS is to form a blob of mass at the end of the ribbon. The centrifugal force pulls it away from Earth, and by careful computer controlled management of its distance and mass, it counteracts the tendency of the ribbon below the GS to fall to Earth. Done just right, the entire space elevator system is stabilized and stays in place, neither falling nor rising. Identical management processes are used to keep the ISS in space, so this is nothing new; we are applying the processes to the space elevator.

Why 100,000 km above Earth? It doesn't have to be this, exactly, but it is a useful distance to maintain a nominal gravity balance. To counteract the tendency of the ribbon between Earth and GEO to fall, requires a long ribbon past GEO, a large mass past GEO, or a mixture of both. For a ribbon with evenly distributed mass, calculations by Edwards demonstrated the ribbon length would be 144,000 km to be stable. But, add mass to the end of the ribbon, and it can be shorter. The greater the mass of the OSS, the shorter the ribbon can be. In fact, as the mass at the OSS changes in operation, the ribbon may be wound in, or out, as required, to maintain balance. In practice, the OSS will range between 70,000 km and 140,000 km from Earth, but in this book we refer mostly to the nominal median distance, 100,000 km. The OSS has been modeled as containing a total mass of about 800 tons, 60% of the 1,400 ton mass of the total system, excluding the mass of the Geo Station which is at the fulcrum of the system.

When it is first emplaced, there will be minimal mass at the end of the ribbon, consisting only of the first crawlers to have dragged ribbon out there. So the ribbon needs to be longer to provide stability. As more mass gets transferred to the ribbon end, so it can be wound in, trading off mass for a shorter ribbon. The final distance selection will be a trade-off of these various factors.

The second role of the OSS lies in acting as a spaceship port for the launching and retrieval of spaceships. Once erected, this is the primary purpose of it.

There is an advantage of a long ribbon between the GS and OSS. The orbit velocity at the OSS increases, the further away it is. When it comes to launching or retrieving spaceships, we can use this extra velocity, adding to, or subtracting from, the relative speed of the spaceship.

A rocket leaving Earth has a velocity around 7 km/sec. The Geo Station, at GEO, has an orbital velocity around 3.1 km/sec, while, at 100,000 km away, the OSS has an orbital velocity around 7.8 km/sec, so from there we can launch spaceships at a similar velocity as a rocket from Earth.

If it turns out CNT is strong enough to stretch to 200,000 km away then the orbital velocity would be around 15 km/sec - fast enough for serious interplanetary travel, even without a rocket boost. This is expanded upon in the Spaceships section.

In theory, the longer the ribbon to the OSS, the greater the velocity boost. The constraint will be the strength of the CNT in the ribbon in tension. There will be a maximum bound limiting how long the ribbon can be; we just don't know what that number is yet.

The third role is to be a space station, just like the GS. All that activity with spaceships will require a large base, a space station, to manage it. But, unlike the GS, the OSS experiences a pseudo gravity. You can think of it as negative gravity if it helps, since this force pulls away from Earth, so the floor in the OSS will seem to be upside down, and Earth will appear to be "up".

The apparent gravity will be low, just 2.2% of gravity at sea level on Earth. Compare that to the Moon, with gravity 16.6% of Earths gravity. OSS gravity will be nominal, but nevertheless real. One impact of this is to put a constraint on the amount of mass at the OSS. We can't measure the upper bound of that constraint yet, but in practice the OSS will have to be smaller than the GS. The OSS will be a smaller, more functional station, dedicated to the job of handling spaceships, managed from the Geo Station.

But for all that, the OSS will be the place to be, where the action is at, dealing with most of the interplanetary space travel.

This is not actually the end of the ribbon. The ribbon continues another few kilometers beyond the Outer Space Station (OSS) to a counterweight and rocket tether at the end of the ribbon, those last few kilometers being regularly paid out or reeled in by computer controlled software that manages the overall ribbon balancing act.

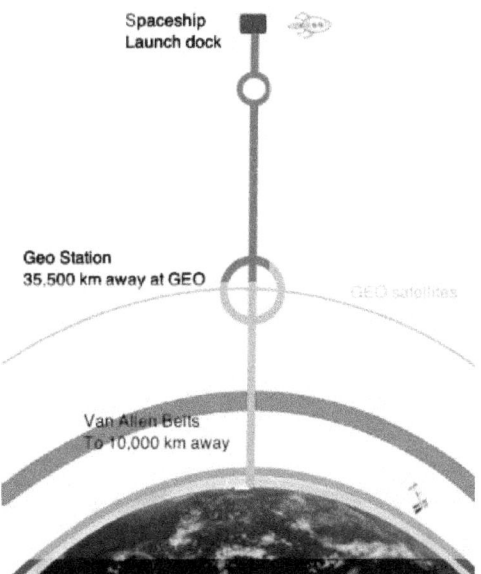

The other difference, at this station, compared with Geo Station, is the number of spaceships. This is the jumping off point for space travel, and a number of spaceships will latch onto the space station, being refueled and provisioned for their journey. This involves the movement of much fuel, which makes it a hazardous place to be. It is not a place where we want to be if an explosion occurs. In fact, the harbor for the spaceships is separated a little way from the OSS. An umbilical cord, about one km in length, extends out from the OSS. Alongside this part of the ribbon are service corridors and fuel lines, and at the end of the ribbon is the harbor where the spaceships are anchored.

This is why the location of the docking modules will be a kilometer or more down-ribbon from the OSS. An explosion at that point would still be dreadful, but it would not be catastrophic in the sense that only the last kilometer of the ribbon would be lost: the rest of the ribbon system would be in place.

You will not travel down there unless you work there or are actually boarding a space-craft. If you do, then you are probably traveling to the Moon (or later on, Mars). See later books in this series for details of travel beyond the OSS to the Moon, Mars and beyond.

13. JOURNEYS BEYOND THE SPACE ELEVATOR

A side trip to the ISS?

Let's deal with a common fallacy here.

Assuming you have your space suit handy, could you "get off" of the cable car at an altitude of 350 km to 400 km or so, and visit the ISS if it is still there? No! Get off of the cable car and you will fall to the ground.

You have altitude and you are in space, but what you don't have is the same orbital velocity as the ISS. Let's go into a bit more detail here about orbital velocity. Earlier, we wrote about the orbital velocity of the ribbon at various altitudes, but we also wrote that it is not in orbit.

In rocketry, the term "orbital velocity" is shorthand for the orbital velocity required at a given altitude in order to remain in orbit without falling back to Earth. Here, we use the term in its wider sense, to mean the rotational speed of an object relative to the core of the Earth. The two interpretations coincide at GEO, which is why that point is a geostationary Earth orbit.

Let's start with you, standing on the Earth. The Earth rotates, and if you are standing still (at sea level, on the Equator, for the sake of this) you are moving at 0.46 km/sec. Another way of putting it, the Earth rotates at 0.46 km/sec, or, you have an orbital velocity of 0.46 km/sec. But you are standing still, so you are motionless with respect to the Earth. It's all relative. Relative to the Earth, you are stationary, relative to space you are orbiting the Earth at sea level.

The ISS orbits the Earth but is not stationary with respect to the Earth. It moves relative to the surface of the Earth, at 7.66 km/sec. Watch the ISS go overhead, and it whizzes by. In fact, at that altitude the ISS requires an orbital velocity of 7.66 km/sec to stay up there. If the ISS velocity was slowed down, so that it didn't move relative to the Earth, it would fall like a stone, because nothing would be holding it up.

Returning to the space elevator ribbon, it is stationary relative to the surface of the Earth, but it possesses an orbital velocity of 0.46 km/sec at sea level, just like you do, standing next to it, rising to about 7.8 km/sec at the end of the ribbon, the Outer Space Station.

For your hypothetical journey to the ISS, at around 400 km altitude, you possess orbital velocity of about 0.8 km/sec, but the ISS whizzes past you at 7.66 km/sec. To hitch a ride, you would need a rocket motor with you, to speed up to match the ISS speed. Having left the cable car it would need to propel you, promptly, to an orbital speed of 40,000 kph to match the ISS.

Until you attain that speed you would be falling so if reaching the ISS was the real aim of the exercise, we would drop you off at a higher level, say about 1,000 km, to allow you time to reach orbit. This would be a cheaper way of reaching the ISS than by launching a rocket from Earth, but a complicated maneuver since you'd still need a rocket to get to the ISS.

Will the ISS still be there by the 2030s? Probably not. LEO will be the home of shoebox satellites with nothing of interest for humans to visit, so you'll be far more interested in carrying on with the journey, out to the Geo Station at GEO.

Connections between Geo Stations

Over time, other space elevators will be built. We have postulated eight of them, distributed around the Earth. This implies eight Geo Stations, too.

In fact, we could build new Geo Stations, without their having a ribbon to Earth, sending components via one of the space elevators, since, being at GEO, they are in a stable orbit. So we can imagine a necklace of Geo Stations in GEO around the Earth, much as we have a necklace of satellites at GEO today. Whether there is any benefit in doing this is a question we can't answer yet, but it's possible.

With many Geo Stations distributed around GEO, it makes sense to enable physical contact between them. It doesn't call for rigid ribbons or roads to do this. Each can be connected by a CNT ribbon, a little loose, not in tension, since a traveling vehicle can loop onto it and use the ribbon as a guide, steering it to the next Station. Besides travel, it will provide a quick and convenient way of a vehicle traveling to a GEO satellite and servicing it.

When traveling around the GEO ring, we don't have the issues of changing orbits or escaping gravity wells. Just like the existing GEO satellites, we can move a spaceship, shuttle, satellite or even an entire Geo Station, around, treating the GEO section of space as a ring road around the Earth, like the M25 in London or the Washington DC Beltway, but less crowded.

A regular service of some kind, traveling around the GEO Beltway, seems likely. This serves a dual purpose. One is the launching, repair and retrieval of GEO satellites. The other is for communication between each Geo Station, for the movement of passengers and goods.

14. "GROUND CONTROL TO MAJOR TONKS"

We will have come a long way since David Bowie sang Space Oddity, with Major Tonks in his tin can. Phillips writes: "I've visited the Smithsonian in Washington and looked in awe at the Apollo 11 Command Module. Same in London where I saw the Apollo 10 module. What's striking is how primitive they were, old rain tanks riveted together. Perhaps we'll relocate Apollo 10 to the Geo Station as a reminder of how far we've gone and one day return Apollo 11 to its landing site on the Moon, where it will be fenced off, and tourists will walk around taking selfies with it."

Does song copyright apply in space? Here's a can of worms. What laws apply on the space elevator and in space, and how are they enforced? Countries, politicians and lawyers will have to address this in detail.

15. CONSTRUCTING AND USING SPACESHIPS

How to travel in space, using the space elevator as the launch platform, is addressed further in the next books in the series, but some salient points are noted here.

The space elevator itself is exciting and will be the biggest advance in technology in the 21st century. But we don't build it for its own sake. The whole point of the space elevator is to provide a cheap road into space, so that we can expand into our solar system, colonizing the Moon, Mars, visiting asteroids and other places. It's the Freeway of the space age.

So, having built it, the next exciting stage is the development of spaceships which launch from it, to the Moon and other places.

At the moment, satellites (we can't even call them spaceships) have weight, mass and size constraints. They have to fit into the nose cone of a rocket. One difference with the space elevator will be seen immediately: we can haul up any number of components into space and assemble them at the Geo Station. So, satellites can be larger, and we can build serious spaceships.

It also makes it easier to build human-rated spaceships with strong radiation shielding. As explained earlier, radiation is a killer issue (no pun intended) in space, and radiation shielding is needed for humans.

Shielding = heavy.

Just the sort of thing the space elevator is good at.

The other advantage is, we can lift fuel into space, rocket fuel. Presently, satellites are sent into space with a bare minimum of fuel. They are pointed towards their target, drift there, using fuel for fine maneuvering only. With the space elevator, we can assemble actual rockets in space, fuel them up, and they can achieve far higher speeds, plus have fuel to decelerate when the destination is reached. This makes it practical to set up regular services to the Moon and Mars. Plus, with shorter journeys, human travelers receive less radiation exposure.

The space elevator is a disruptive technology. The rocket industry will see it as a threat, initially, but like all disruptions, they will learn to work with it. If the demand for heavy lift rockets from the surface of Earth reduces or ceases, there will be a rapid expansion in demand for rockets traveling in space, powering our new spaceships. Even with the space elevator, our spaceships need rocket power to get to the Moon, Mars or anywhere else. In time, we will be building far more rockets than we are today. The difference is, they'll be taken up to the GS, assembled there, and perform as reusable rockets in space.

Elon Musk talks of plans to use SpaceX to take humans to Mars in the 2020s - possibly one-way. Full marks for the adventurous spirit, and it will be a spectacle if it happens. But if the original Apollo mission to the Moon was risky, this is doubly so.

If we wait until the 2030s for a first human Mars mission, the parameters change and we can describe a safer mission, perhaps many of them.

Using the Geo Station to assemble spaceships there, many of them, like a line of truckers on the road, we'll send out a steady stream of space trucks to Mars, putting a supply of goods into orbit around Mars, also landing some, contributing the bits and pieces of a first Mars colony.

Having built the GS and OSS, we could build another large space station, and using rockets, send that off to Mars, to orbit it. This Mars Space Station (MSS) would be ready and waiting for the first human mission to arrive. Of course, we'll have sent supplies of air, fuel and water too.

This way, when the first human Mars mission sets off to Mars, it will have everything ready and waiting. In addition to the spaceship, it can be accompanied on the journey with a spare spaceship, and space trucks, with plenty of backup in case anything goes wrong. Plus, it's far more likely these humans can come back again.

This is the way to travel to Mars. Allowing for the two-year cycle of Martian oppositions, we aren't limited to just one spaceship. In fact, this will be the start of scheduled runs to Mars of both spaceships and space trucks, which also means regular returns.

First off, we'll need rockets to land on Mars, but with rockets and plenty of fuel, we can do it the easy way, instead of using hoppers or bouncing balls. We have plans to erect a space elevator on Mars, covered in a later book in this series. Once the Martian space elevator is erected, two-way travel between Earth and Mars is made simple.

We've described, briefly, travel to Mars above, but the same principles apply to Moon travel. With a space elevator set up on the Moon (also covered in a later book in the series), the relatively short hop from the space elevator to the Moon elevator makes it feasible to have a daily service to the Moon, perhaps even several journeys a day. Now we're talking! We are starting to see space travel evolving, the way air travel evolved in the last century. This is how we make space travel as routine as air travel.

The scenario above will probably be owned and managed by a country, a multinational group, or a large corporation, because of the costs involved. As more nations get their own space elevator, it is our hope they will cooperate in space exploration, though competition is likely.

What about individuals? Many of us will be impatient to get into space. Elon Musk certainly is. We can imagine high net worth individuals ordering their own spaceship to be assembled at the GS, just like they have a personal jet today. There will be competition to achieve "firsts".

Even as we get the travel systems for the Moon and Mars set up, some high net worth individuals will try and leverage off of it, to be the first to get to Mars, say. Other objectives, or firsts, can be imagined: the first to land on Phobos and build a base there; the first human to orbit Venus or land on Mercury, or visit an asteroid. We can send out mining probes to the asteroids by the dozen.

Someone will no doubt try to set space travel records. Going beyond Mars, even to the asteroid belt, becomes a long journey of years. But, because it's there, someone may load up their own spaceship, to be the first to reach one of the moons of Jupiter or Saturn. They may never come back from such a mission, but, for an older high net worth person, they may view it as a fitting, history making, achievement with which to end their life. Humans always did do crazy things.

16. FLY ME TO THE MOON

The logistics of space travel, the distances, the relative velocities of the Moon and planets, are addressed in detail in later books in this series. The series also addresses the next milestones, building space elevators on the Moon, Mars, and other places.

Other places? It's a pity our solar system isn't populated with a dozen Earth-like planets, all green and full of life, waiting for us. Then we'd be far more energetic about space travel. The sad reality is, our solar system is a tough place for humans.

Mars is an obvious destination, but even that is a dead, rocky planet half the size of Earth, with a surface ruined by perchlorate. Perchlorate is nasty stuff, used in munitions, fireworks, explosives, matches, signal flares, fertilizers, chlorine cleaners, and pool chlorination chemicals. Movie images of a lone spaceman growing potatoes in Martian soil are a long way from reality. The Dead Sea in Palestine is a better analogy.

Every other planet or moon is even less attractive, scattered far and wide, with the emphasis on "far". But we have to start somewhere.

Our first objective will be the Moon, so here we wrap up with a primer on travel to the Moon.

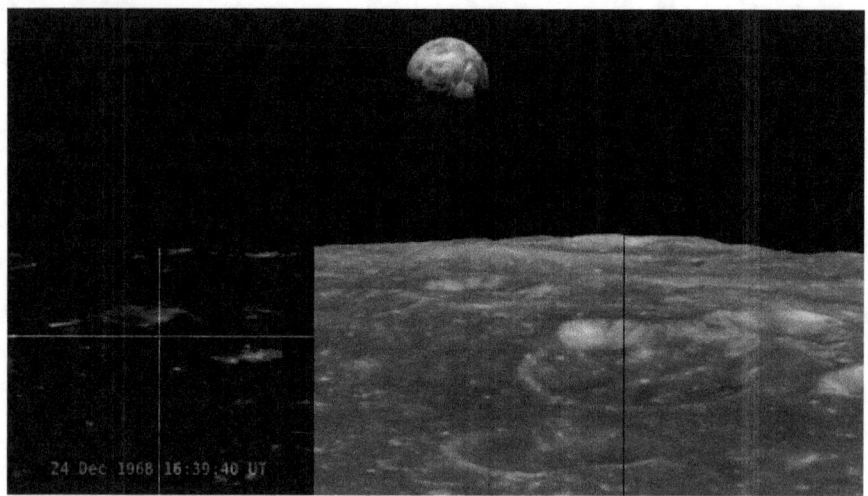

Earth-rise from the Moon. Image source NASA
Scientific Visualization Studio

See book five in this series for a detailed look at Moon coloniza-
tion.

We cannot build a space elevator from the Earth to the Moon, nor
to any planet. In space, they all keep moving around. The purpose
of the space elevator is to get out of gravity wells, whether it be
the Earth, Moon, Mars or anyplace else. Having escaped from a
planetary gravity well, we use rockets or some powered vehicle
to travel elsewhere.

Traveling to the Moon, even from the Outer Space Station, requires higher kinetic energy trajectories. Think of it as the Moon
traveling faster, so we need more speed to catch up.

The Moon orbits around the Earth. Also, the Earth rotates, once
every day. So, relative to an observer on Earth, the Moon changes
position every day. Take a look at the Moon tonight. Go out to-
morrow night at the same time. It will have moved East, by 13.2°,
about the width of a hand at arm's length. Imagine the space elevator pointing directly at the Moon. It is advancing East, so, as the
Earth rotates, the space elevator moves away from it, until it's
on the other side of the Earth, half a day later. It takes 24 hours

49 minutes, on average, before the Moon is directly overhead of the space elevator again. Also, to complicate things further, the Moon does not orbit around the Earth directly above the Equator. The Moon's orbit is tilted by 6.7° relative to the Equator.

If we launch a spaceship from the space elevator, heading for the Moon, there is a journey time to account for. It took Apollo four days to reach the Moon. From the space elevator, the journey will take less than a day, but it depends on, firstly, whether it departs from the GS or the OSS, whether rocket power is used, and whether it is to land on the Moon, orbit the Moon, or arrive at a Moon space elevator.

All these journeys have one thing in common: in orbital terms, they start off ahead of the Moon and appear to slide backwards, getting into position for the Moon just as it arrives at the correct rendezvous point. This is referred to as allowing for the retardation of the Moon in its relative orbit.

The orbital velocity, around Earth, of the Geo Station is 3.1 km/sec, and for the OSS 7.8 km/sec. Relative to the Earth (and the space elevator), the orbital velocity of the Moon is 30.1 km/sec. (The Moon orbits the Earth at about 1 km/sec, a leisurely pace. However, that is relative to an Earth that is not rotating. Our Earth also rotates once each day, so compared to a fixed point on the surface of the Earth, the Moon appears to have a velocity of 30.1 km/sec.)

What this means is, for a spaceship leaving the space elevator, the Moon is catching up with it, at a high relative speed. So, not only has a spaceship to travel outwards, into the orbital plane of the Moon, it also has to speed up, to match the Moon at 30.1 km/sec. This means a relative velocity adjustment of 22.3 km/sec. For the return trip, the opposite applies, as the spaceship has to shed velocity to dock with the OSS on the Earth Elevator. Plus, depending on what time of the Lunar month it is, we have to allow for the orbital inclination of the Moon, 6.7°, which may mean additional travel north or south, to adjust orbits.

A space elevator on the Moon, assuming it to be 200,000 km in length, would have a relative orbital velocity around 15 km/sec at the equivalent OSS, the Moon Space Station (MSS) on the Moon Elevator.

With plenty of fuel and rockets, this range of orbital maneuvering is well within a manageable range. We already do this from Earth, for satellites heading to the Moon, which have a relative velocity adjustment of 30.1 m/sec to achieve. Once we've built the Moon space elevator, the relative velocity adjustment can be reduced. For a Moon Elevator hanging down towards Earth, the Earth end of the Moon Elevator has a lower relative velocity.

How low depends on how strong CNT turns out to be. Theoretically, on computer we can model a system where the OSS is 200,000 km from Earth, and the Moon Elevator extends 200,000 km to Earth. The figures are approximate, but at this distance, the two elevator ribbons would pass close to each other, in which case the relative velocities of the end of each ribbon would be less than 1 m/sec, making a transfer easier, though still rocket powered. Presently, it seems unlikely this will be achieved, but it's not impossible.

A more feasible model compares a spaceship leaving the OSS on the Earth Elevator, 100,000 km away from Earth, with an orbital velocity of 7.8 km/sec, and traveling to a Moon Space Station on the end of a Moon Elevator extending 100,000 km down from the Moon, which would have a relative velocity of 21.8 km/sec. This narrows down the delta, the velocity adjustment d(v) to 14 km/sec, well within our existing capabilities. So, depending on the actual lengths of the respective space elevators on Earth and the Moon, when constructed, the d(v) will lie somewhere between 1 km/sec and 14 km/sec.

17. LAUNCHING PAYLOADS BEYOND THE SPACE ELEVATOR

Lastly, a comment about launching payloads away from the Geo Station. Once we have gotten our payloads up to GEO, they don't just stay there. We want to deploy them elsewhere.

Let's say you have a satellite that is to be deployed at GEO but 90° to the Geo Station, along the GEO line. Although GEO is a notional circle around the Earth, from the point of view of the Geo Station, it's a straight line in either direction. So, a satellite needs propulsion to move away from Geo Station at a speed that keeps it at GEO.

The problem for us is twofold. We don't want a satellite to use propellant propulsion that blasts the Geo Station, so we have a procedure that simply "nudges" the payload away from Geo Station, delaying any propulsion until it is a safe distance away.

The other problem is that the payload has mass. When it propels away from Geo Station, the station has an equal opposite reaction in the other direction. Left to itself, that reaction will feed down the ribbon and the station will end up oscillating with an amplitude around the equilibrium position. Depending how much amplitude, we can use propellant to dampen it. But also, we have cars coming up the ribbon, transferring their own amplitude to the ribbon system. The ideal is to have the car movements and payload ejections balancing each other.

When a payload is ejected, the Geo Station also loses mass. The effect will be a function of the mass of the ejected object over the remaining mass of the station. Over time, we'll build up a very large mass at the station. Of course, it gains mass every time a car arrives with payload, so in effect we are carrying out a spacial juggling act of mass.

18. THE THRILLING CONCLUSION

What will be the next major advance? Will we discover how to build a warp drive, Star Trek style, or discover wormholes in space? As humans, we live in hope, but sadly, it may be that such things don't exist. In past decades, we've achieved great advances in our understanding of the universe, but the reality of what we see out there is this: everything in the universe seems to be explainable by natural causes, or at least, what our physicists understand them to be. So far, we've not found any aliens, nor have we seen anything requiring an advanced technological civilization to explain it. No tell-tale tracks of warp drives, Dyson spheres, no communications traffic, no signals from passing spaceships, in fact, nothing suggesting any technology beyond what we are building.

It may be that the space elevator is the last and best advance in technology for space travel. It would be sad if so, for it means, at best, humans probably stuck exploring our solar system, an unfriendly and unforgiving place. We have plans for colonies on the Moon and Mars, but there'll be no wormholes jumping to Pluto or anywhere. If traveling around our solar system looks difficult, then interstellar travel looks even harder, doing it the slow way, sending spaceships on a journey of centuries just to get to our nearest neighbor, Proxima Centauri, or Barnards Star, and still no nearer to answering the age old question, what's it all about?

Perhaps those advances, and interstellar space travel, are the stuff for people of the 22nd century. People of the 20th century, wars aside, lived through an incredible time, when people went from living in their own little towns or villages, to using aircraft to fly around the world, taking holidays in Hawaii, visiting Antarctica.

Our world has changed like it has never changed before - from horsepower to space travel in a century. Our 21st century is going to witness the next wave of expansion. Using space elevators, we will expand into our solar system. Our grandchildren may view space travel the way we've accepted air travel today. By the end of the 21st century, it should be a wonderful sight.

Space Elevator Operations

Book 3 of the Space Elevator 2020 series

Published 2020

Linda J. Phillips

The Space Elevator 2020 series
Book Three:
Space Elevator Operations

Publisher contact: info@21stcentury.space
FaceBook www.facebook.com/lindyjaniceAuthor
Twitter @_lindaphillips
Amazon author page https://www.amazon.com/author/linda-janicephillips
Web 21stcentury.space
Web links utilized in this publication were correct at the time of writing, but they can change over time.

Published by Linda Phillips